Made in United States
North Haven, CT
06 November 2026

21876158R00070